I0075106

Grand Unification

Can the "theory of everything" be this simple?   by L.N. Smith

The lights dim in the planetarium, and a hush falls over the crowd. The show is set to begin.

Through the darkness, the sound of a ticking clock marks time like a metronome . . . faintly at first, but then louder and louder, until it clangs like a rail spike being driven by a sledgehammer.

The cadence slows; the trailing echoes linger ever longer. Then, nothing. Time seems to stall.

KABLAM! the void is shattered. I'm blinded by the glare off the ceiling, and a shockwave rumbles through me.

When the initial blast subsides, I turn my eyes domeward, where the image of a fireball expands rapidly from the center. It engulfs the ceiling with a roar, then all is dark and quiet again.

The pop of a microphone breaks the silence, and the audience reacts with a knee-jerk groan.

"Is this on?" comes a raspy voice through a speaker system. "Testing . . . Testing . . ."

The man speaking is nowhere to be seen, but his voice is coarse and wavery, like that of an old cowboy. He blows into the microphone and taps it. "Hello?" he says. "Oops . . . I mean . . . Welcome! Welcome to the Admiral's Planetarium. I'm your unearthly host, Norbert Binnacle."

Norbert clears his throat. "Your ten-minute visit to this

planetarium will be an exciting one. I'm going to explain the workings of your universe and the creation of your small blue planet. But before I begin, I should tell you a little about myself.

"I'm an instructor for Admiral Boom's *Satellite Education Program*. I'm sure you know the Admiral, though probably by a different name. The good Admiral is my boss and mentor. He trained me for this job, and even though I finished at the bottom of my class, it turns out I'm the most requested teacher on staff. I credit my success to my inability to actually master the course material. You see, I couldn't picture parallel universes . . . or nine physical dimensions . . . or antimatter . . . or virtual particles . . . or any of the other highly abstract concepts my classmates claimed to grasp. And you know what? I quickly discovered that my audiences couldn't grasp them either.

"So, you won't believe what I did—ah, the audacity. I came up with my own design for the universe. It was kind of a fluke, really; I was just trying to understand my notes, when a sensible solution hit me. At last, I could dismiss the troublesome notions of parallel universes and nine physical dimensions, and I could finally embrace antimatter and virtual particles as old friends. My new idea even mended the great schism of science—the incompatibility of *general relativity* with *quantum mechanics*. Whew! What a relief.

"But let me be the first one to admit that my lecture today might not be entirely accurate. You see, I haven't done the math." Norbert laughs. "But that shouldn't bother *you* guys much. Your human equations are a wreck, plagued by pesky terms that stray to infinity. Impossible! Allow me to explain.

"On glorious mornings, Admiral Boom likes to shake things up a bit. *No one sleeps today,* he'll say. *Mr. Binnacle! Put a double charge in the time gun.* The Admiral has never asked me

© 2004 BY L.N. SMITH

to load an *infinite* charge into the gun. How would I do that?

"Thus, without any math to muddy the works, I was free to dream up any universe I pleased.

"So, make yourselves comfortable . . . especially you science-minded folks. You in particular will find ol' Bert's universe to be a breath of fresh air . . . a helping hand . . . a welcome realignment to your proverbial undergarments. In the course of these ten minutes, I'm going to put the flow back in your step, rebalance your lopsided particles, and pry your energy out of a bundle. The first order of business is to rewind the film you just saw.

"That explosive introduction was the Hollywood version of the Big Bang—you know, the start of the universe. It's a crowd-pleaser for sure, but somewhat misleading. And besides, it sidesteps the most fundamental question of all: What is exploding? —What is the universe actually made of?

"The answer to this question is the key to understanding everything else. So get ready, 'cause this blundering student of the cosmos is about to tell you. Please direct your attention to the ceiling."

Projected overhead is a picture of a solar system, with a sun at the center and several planets orbiting around it.

"This solar system is the perfect way to illustrate the atom," says Norbert. He hesitates. "Hmm. I suppose some of you may have forgotten your high school science classes? Let me review."

The video on the ceiling pauses.

"Atoms are the tiny bits that make up just about everything you know, like rocks, trees, and people. And did I mention they're tiny? —At one time, they were the very definition of tiny. Atoms are so small that a single human brain cell contains about a million billion of them. And each atom is built like a solar system."

The picture on the ceiling springs back to life, but this time with a new look. The sun has transformed into a clump of wiggling soap bubbles, and each planet has dissolved into a filmy sphere around it.

"The nucleus is the heavy piece at the center, like the sun, and the electrons are the wispy orbiting planets.

"This childish representation is more accurate than you might think. Not only does it have the right shape, shake, and shimmy, but also, its milky bubbles all arise from the same constituent material, just like the particles of a real atom do. That's right: all the pieces of an atom are spun from the same stuff. And even the many particles *not* found in atoms are spun from this stuff too.

"So, where can you go to meet the lively cast of subatomic particles spun from this curious material? Why, the Particle Zoo, of course!

"Ah, the infamous Particle Zoo: it houses several dozen critters. Among the best known are the gluons, the muons, and the neutrinos. But the biggest draw to the zoo by far is the popular quark family, with its crazy shade-wearing members, named Up, Down, Top, Bottom, Strange and Charm. The characters of the Particle Zoo come in many sizes, yet all of them are giants relative to the fundamental material that makes them. Just as the animals of a real zoo are built from tiny cells, the particles of the Particle Zoo are built from tiny . . .

"But I'm getting ahead of myself. The smallest particle in the universe is something that sails around like a comet . . . never slowing down and never speeding up. It never collides with another of its kind, yet it maps the routes traveled by all. This smallest particle is the basic building block for every-thing in the universe. It forms the long-sought ether—the

© 2004 BY L.N. SMITH

fluid of the cosmos—and I've named it the ethercomet, or e-comet for short.

"E-comets explain a lot of things your scientists cannot. E-comets explain why particles can be converted into 'pure energy' and why 'pure energy' can be condensed into particles. They explain why the miniature world of atoms is ruled by probabilities rather than absolutes. They explain why light has the wacky nature of behaving as both a wave and a particle. They explain why precise clocks can't agree on the time. And finally, e-comets explain why the universe is so many times more massive than all of its stars and galaxies combined. The truth is, space is not a void: the dark matter and dark energy that fill the cosmos—yet defy scientific explanation—are, in fact, one and the same thing. They are e-comets cruising solo.

"So, what's an e-comet? The answer may surprise you.

"An e-comet is not actually made of anything: it's not chunky or hard like a rock. It's . . . well . . . Allow me to illustrate with a piece of fabric."

On the ceiling, a rectangular cloth appears with a grid printed on it. A threaded needle weaves its way along one edge.

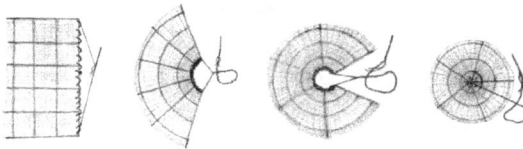

"Imagine outer space as this piece of fabric. If we pull the thread tight, we cause an entire side to collapse into a single point. The ends of the fabric bend around until they meet and form a circle.

"An e-comet is like this fabric. It's a place where space is bent around and closed tight into a point, as though pulled that way by a drawstring. There's nothing solid to find in the middle."

Norbert takes a breath.

"Ah, but I hear a complaint: the fabric is only two-dimensional; space has three. That's an easy one. Watch carefully the *Throw-pillow of Science*."

A hologram appears overhead, showing a disk-shaped pillow with a needle weaving its way across one whole face.

"When the thread is pulled tight, an entire side of the pillow collapses into a single point, bending the rest of it around into a sphere. This is the shape of an e-comet.

"BUT!" The hologram pops. "Working in three-dimensions complicates the graphics. We'll stick with flat pictures for the rest of the presentation. Back to the fabric.

"Now, let's hide the fabric and instead show our little e-comet as a single point of light. And let's give him a friend by adding another comet to the display."

From the edge of the ceiling, a second e-comet makes its entrance. Its arrival sets the first e-comet in motion, and the two move toward one another. But their flight paths don't meet. Rather, the two e-comets end up tumbling around each other in a tight circle.

"Watch as a third e-comet enters the ceiling."

Again, the new arrival sets the others in motion toward it, but this time they cross the ceiling at different speeds. The tumbling pair is slower to move than the single e-comet. The new arrival meets the pair, but it doesn't enter their rotation. Instead, it slingshots around them like a comet from outer space, swinging way out before looping back in for another close pass. The three e-comets repeat this gyrating pattern around the ceiling.

A fourth comet enters, then a fifth. Each new arrival

© 2004 BY L.N. SMITH

affects the flight paths of the others, kinking and bending their trajectories.

Norbert calls our attention to the fact that the comets never leave the ceiling. They simply swing out and then return for another pass at the center.

"The universe is contained," he says, "—contained within the region traveled by its e-comets. None of them escape."

Reflecting on Norbert's last words, I let my eyes lose focus on the ceiling. The spots of light blur and blend together, forming a circle that grows and shrinks in endless cycles of expansion and contraction.

"Let's flood the ceiling with e-comets," announces Norbert. "Then you'll see something *really* spectacular."

The existing e-comets shrink to the center, as the picture widens to include thousands more.

"Watch the patterns emerge where clusters of e-comets travel in tight groups—chasing, looping, and twirling around one another. These piled-up strands of e-comets look like helpings of spaghetti . . . or the inner-windings of a baseball . . . or perhaps the hyper-creations of a Spirograph. Who needs nine physical dimensions when there are these messes to untangle?

"You humans call these clusters *particles*—or *matter*—and they constitute the many critters of your Particle Zoo.

"Notice that some of the clumps are unstable and disperse as soon as they appear. Your scientists refer to these as *virtual particles* because the clumps don't last long enough to be observed. You might then ask: so how are they known to exist if they can't be observed? It turns out that these virtual particles cause measurable effects during experiments.

"Some clumps of e-comets last longer. But as you can see, they can be destroyed in the blink of an eye. When a clump encounters a second clump that is its own mirror image, the result is immediate annihilation of both, as e-comets scatter in all directions. Perhaps you know the look. If you've ever paddled a canoe, you may have noticed the two whirlpools that can form behind the paddle and which spin in opposite directions. If these whirlpools collide, they cancel each other out in a flash of smooth water.

"Your scientists have adopted unfortunate names for these pairs of opposing clumps. The terms *matter* and *antimatter* suggest very different things, when in fact the clumps they describe are nearly identical. Perhaps a name change is in order—something more intuitive like *mirror-matter* or *counter-particles*.

"The mention of antimatter brings up some interesting questions. Why is there so much more matter in the universe than antimatter? Shouldn't the amounts be roughly the same? How could the great canoe paddle of the cosmos create one whirlpool and not the other? Why are particles so out of balance? Where's the symmetry?

"The answers to these questions are simple: there isn't; they are; it didn't; they're not; and it's there.

"The universe is divided into regions of matter and antimatter. There is, however, no way to tell them apart with telescopes or other long-range sensing devices, because these regions send out identical signals. The best way to confirm an opposing region is to go there . . . or to have it come to *you*.

"But, for as long as the universe continues to expand, such meetings will be few and far between. Only when the universe one day collapses will the sky become ablaze with the fires of annihilation, as whole galaxies swallow each other in the death throes of a universe on the verge of rebirth."

Norbert sighs and sips a drink. "Back to the spaghetti-like clumps of e-comets.

© 2004 BY L.N. SMITH

"Even the piles that manage to hang together are not entirely stable. They are endlessly absorbing and shedding e-comets, and sometimes converting themselves into other kinds of 'particles.' These fluid transformations happen on such a small scale that the tools of human science cannot accurately track them. Therefore, scientists must resort to probabilities when predicting outcomes, and such uncertainties have turned your scientists into philosophers.

"Some scientists speculate that the Admiral is rolling the dice at every turn, enjoying the universe as one giant crapshoot. Others suggest that the dice are loaded, and that the Admiral knows every outcome ahead of time. And a few scientists have even engaged the fanciful, suggesting that every roll of the dice produces every possible outcome, creating a new parallel universe to house it. Awesome.

"There is, however, a simpler and more plausible explanation for the use of probabilities. It's a well-known fact that card games are governed by chance when the cards are shuffled and dealt face down. But show the cards to the players in advance, and all probabilities disappear—all outcomes become known. E-comets are the cosmic cards dealt face down. The outcomes they produce are relegated to chance because each e-comet is too small to observe.

"Now, there's another feature of these e-comet clumps I need to point out. Notice how some of the piles cause patterns to form around them. And see how those patterns intensify when two such clumps get near each other, sometimes pushing them apart and at other times bringing them together. These are the patterns you humans call *electric charge*—one of your Big Four of natural forces. But electric charge is not truly a force. It is simply the e-comets going about their business, following their natural paths through space. There is no tugging or shoving going on at the e-comet level. The force you feel is simply the large-scale effect of these e-comet movements.

"Likewise, *gravity* is another of your Big Four that is not really a force. Remember the cloth we pulled into a circle? It demonstrated that space is not straight, but rather, curved. How else do you account for oceans that hug a spherical planet or for a moon that stays in orbit without ropes or motors?

"And so it goes for the last two forces, named *strong* and *weak*. These occur at the center of atoms—in the nucleus to be exact. But they are nothing more than e-comets roiling about, like bees in a beehive. Smash the beehive, and you release the bees. Welcome to the atomic bomb."

There's a bright flash at the center of the ceiling, and the e-comets ripple outward like waves across a pond.

"This brings us to the concept of light. And for this explanation, we'll be heading to the beach."

The scene overhead changes to an aerial view of a sunny coastline, where a retreating tide has stranded some water high on the beach. The water in this isolated pool sits perfectly still, its surface like glass.

"The water in this tidal pool looks calm, but any good scientist will tell you that it's swarming with tiny particles. If I disrupt the water, these fast-moving particles register the disturbance by sending out a wave. Air is the same way: it is teeming with atoms in chaotic flight. If you disturb the air, a sound wave results, peeling across the sky at nearly a quarter-mile per second. The universe is the same way: it is teeming with e-comets that sling around each other, registering disturbances with light waves that top out at 186,000 miles per second.

"Water, air, and e-comets—it doesn't matter which medium you choose—each one produces waves well-versed

© 2004 BY L.N. SMITH

in the usual dance steps, like Diffraction, Interference, and the Doppler Shift. But there are other dance steps too, ones that involve a larger partner. When it comes to e-comets, it is atoms who like to join them on the dance floor. We'll be using water and pool toys for our demonstration today."

Four plastic rings of different configurations appear suddenly, floating in the tidal pool without having disturbed the water.

"Watch what happens when I agitate the pool," says Norbert. "Keep your eyes on the rings."

A scattering of small stones plunk into the water, sending out waves in all directions.

"Notice at the first ring, the waves bounce off the outer rim . . . 'reflected' you might say. At the second ring, the waves pass easily across . . . like light through a window. At the third ring, the waves get trapped inside, causing the ring to shuffle around . . . like bare feet on hot charcoal. And at the fourth and final ring, the incoming waves are actually harmonized and recast as new waves . . . like the tuning of an orchestra.

"These pool toys illustrate what happens when waves of e-comets wash up against atoms. They are either turned away, permitted through, or held inside as heat until released.

"Such is the wave nature of light. But remember, light acts like a particle too.

"Of course it does! It must!

"Light is made of e-comets—the same stuff that makes matter. Try standing in the ocean and letting the surf roll against you; you'll feel the water pushing you backward. Light waves exert a similar pressure, calculated at $4\frac{1}{2}$ pounds of sunshine against the earth.

"So why is light such a puzzler for your scientists to understand? Why is its dual nature to act as both a wave and a particle so difficult for them to reconcile, especially when water waves and sound waves, which also have this dual nature, cause them no such trouble? The answer, it turns out, has to do with the human tools for studying light.

"When humans measure the energy of light, they always find it traveling in bundles of a certain size. It's as if one must try to imagine the ocean waves traveling in zipper-lock bags. But there's an error here, and it involves the word *travel*. Light waves, just like ocean waves and sound waves, don't travel in packets; rather, the bundling is merely a circumstance of the way the arrival is observed. Watch the youngster in the shallows just off shore."

A small boy grasps the rim of a pail, pressing it firmly against the sandy bottom between his feet. The next wave has nearly reached him.

"This child's pail is set to catch the incoming wave," says Norbert. "It's not important how big the wave is: if it gets over the top of the pail, it will fill it. But the pail is a clumsy tool for measuring the wave. It might easily lead the boy to conclude that waves travel in quarts, since that's the amount of water he catches and dumps each time.

"Atoms are much like this pail. They catch and dump relatively fixed numbers of e-comets, tricking their observers into believing that light travels in packets of a uniform size.

"The year was 1905 when the young patent clerk, 26-year-old Albert Einstein, had the notion to package your light waves and solve the problems of his day. His idea single-handedly propelled science forward with this practical, yet inconceivable, conclusion. I hope the e-comet ends your century of confusion."

The camera zooms in on the boy's pail, where the words PHOTON CATCHER are stenciled on its side.

Beyond the boy, someone else wades in the water.

© 2004 BY L.N. SMITH

"The boy's father has brought his own toy to the beach today."

A man works beside a giant floating ring of 55-gallon drums laid on their sides and fastened end-to-end to form a square. Anchors and chains keep the rig from drifting away in the turbulent surf. At the center of the ring, where the water is relatively calm, a stick protrudes from the sand to measure the water's depth. The words GRAVITON CATCHER are painted across the barrels.

"The boy's father also likes to study the ocean. He plans to measure the gentle rise and fall of the tides by noting the watermarks on the stick. His floating ring of barrels is the heavy barrier he needs to block the constant wave action that would otherwise disrupt his readings.

"The ocean tides represent waves of e-comets quite different from those portrayed by the rolling surf. The tides symbolize huge sweeps of e-comets cast out by the motions of massive heavenly bodies. These tides are incorrectly termed *gravity waves,* and their best producers are collapsed stars trapped together in speedy orbits. Such rotating pairs spray e-comets like water from a pinwheel.

"Light waves and gravity waves: they are mere swells in the ocean of e-comets, riding along at the speed of light.

"And that brings me to my next point. Why do you humans presume that the speed of light is the universe's top speed? Consider sound waves, for example: they

---

GRAVITY IS NOT DELIVERED

Gravity is not a force, nor is it transmitted at the speed of light. Gravity is simply an effect brought on by the curvature of space—a curvature defined at every instant by every e-comet in every location across the cosmos.

travel at about half the speed of their constituent carrier. Why shouldn't it be the same for light? Why can't light be the lazy wave through its medium of zippy particles? Well, I'm here to tell you that it can be and it is. Light does not hold the title of *Fastest in the Universe.* If you want to discover real speed, try parting the sea of e-comets and sending one through.

"But you humans can't yet part the sea of e-comets. The best you can do so far is to muscle through it.

"In your expensive colliders and accelerators, you drive your particles closer and closer to the speed of light without ever reaching it. You plow through an ever-growing resistance of e-comets, vainly spending more and more energy to gain less and less speed. Your particles grow massive in the stackup, and your equations blow up, crying INFINITE ENERGY! INFINITE MASS! Your struggle is a classic one, as you attempt to recreate the conditions of the early universe . . . a time when e-comets traveled closer together . . . a time when space was much denser.

"That's right, I said *denser.* Space has density, didn't you know? It's right there on our picture of the e-comet."

The image of the fabric returns to the ceiling.

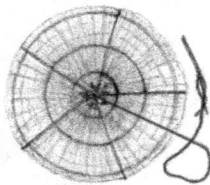

"Did you happen to notice that none of the material was actually cut away when we pulled the rectangle into a circle? It simply became bunched at the center, cramming tighter and tighter together.

"For the e-comet, it is space that undergoes this compression,

© 2004 BY L.N. SMITH

becoming more and more dense toward the center. It is this density that is responsible for some astounding effects . . . well, density and something else.

"The other feature I'm referring to is the e-comet's constant speed through space. Remember that the e-comet never speeds up and never slows down. It's no joke. However, the speed I'm talking about is not the kind you're used to; it's not measured in miles per hour or meters per second. Rather, the e-comet's speed is measured in units of space per ABSOLUTE COSMIC TIME."

Norbert's last words—*absolute cosmic time*—reverberate through the auditorium.

"That's right. There's a master clock, and it ain't yours. While you may set your watch by Greenwich . . . Greenwich takes its time from the Admiral. It's the Admiral who holds the master clock. Your clocks all tick at different speeds.

"Now, I know it might seem strange to think of clocks in this way. Shouldn't all good quality timepieces tick at exactly the same rate no matter what? The answer is *no*. A high-precision clock on the ground ticks ever-so-slightly slower than an identical clock hovering overhead in outer space. What causes this difference? Perhaps you've guessed that the answer involves two things: the constant speed of e-comets and the varying density of space.

"So here's my answer, explained in terms of rabbits and carrots. If two hungry rabbits eat carrots side-by-side in matching rows of a garden, they'll finish eating at the same time. But what if one row is planted more densely, having twice as many carrots? Then, when both rabbits chew at the same speed, one rabbit will finish his row twice as fast as the other.

"Yet I find that some people don't like carrots . . . or rabbits. Let me try another approach. Back to the beach."

This time there are two tidal pools, and one of them lies in the shadows of a row of palm trees. Drawn on the bottom of each pool is the face of a clock showing numbers but no hands.

"Both ponds hold sea water, but there's a difference between them. One pool—the *fast pool*—sits in the sunshine and is warmer than the other. Its particles of water move more quickly and reside farther apart than those in the shaded pool.

"The shaded pool, then, is the cool pool, also known as the *slow pool*. Its water particles move more sluggishly and are tightly packed together.

"I'm exaggerating when I say that the cool pool is twice as dense as the warm pool, but exaggeration helps my example. Watch what happens when I add identical windup submarines to the basins."

An odd-looking toy appears in each body of water. It has the shape of a submarine, but its hull is a mesh of wires. Water floods the interior, where the inner workings sit in plain view. They include a plastic pilot, a rubber band, and a propeller. Each submarine begins its journey above the twelve o'clock shown on the bottom of its respective pool.

"Recall that I said the subs are identical, but note that one of them looks larger than the other. In the warm pool, where the water is less dense, the submarine and its driver have spread out . . . and I mean literally: they're swollen. In the cool pool the effect is just the opposite: the denser water has compressed the sub and its driver, making them smaller. Consequently, if each driver grabs a ruler from his glove compartment and measures his own submarine, each will get the same answer, because the rulers have changed size accordingly.

"Note also that the larger submarine is way ahead of the smaller one. The sub in the fast pool is almost at six o'clock, with its rubber band nearly unwound, while the sub in the slow pool is only at three o'clock, with its rubber band only

© 2004 BY L.N. SMITH

half unwound. The denser water is slowing everything down. Even the waves move half as fast in the slow pool.

"Although the passage of time feels the same to both pilots, their years tick away at different speeds. The pilot in the fast pool reaches old age well before the pilot in the slow pool.

"So where can you go to enjoy denser space? In other words, where can you go to slow your clock and reduce your dress size? The nearest place to go is the sun, but you'll certainly be disappointed there because its density is barely more than that of Earth. You'll have to go much farther away to get a

---

MIXED SIGNALS

Connecting the slow and fast tidal pools by way of a narrow channel permits waves to travel between them. Waves originating in the slow pool accelerate and stretch as they move through this channel. Waves from the fast pool do just the opposite: they slow and stack up as they enter the denser water. Thus, the remote waves received by the two submarine drivers tell very different stories.

The driver in the fast pool detects waves that are radically *red-shifted*. These low-frequency, low-energy waves mean that the remote submarine is either sitting in denser water or racing away at high speed, or perhaps doing a combination of both. The slow-pool driver, on the other hand, experiences waves that are radically *blue-shifted*. These high-frequency, high-energy waves indicate that the remote sub is either sitting in low-density water or racing closer, or both.

Gravitational shift is the name for this effect when caused by the density of space. *Doppler shift* is the name when the effect is caused by motion.

dramatic effect. The place to go is a collapsed star—called a black hole—where even the light waves slow to a crawl.

"However, if such a drastic relocation doesn't appeal to you, there is something else you can try—an alternative almost as good, and some say better. It, too, slows your clock, but it doesn't change your dress size entirely. Rather, it simply flattens your butt and belly. Here's how it works.

"Recall that your butt and belly . . . and, in fact, *all* of your body parts . . . are nothing more than generous servings of e-comets cruising in tight formation, and that the size of these servings depends on the density of space. But what if you could somehow change the density of space in one direction only? What would happen to your piles of e-comets *then*? The first answer is *they would flatten.*

"As the circulating e-comets spend some of their cruising time in denser space, the distance they travel in that direction shortens. The result is a flattening of the pile.

"To explain the second effect—the slow-down of the e-comet circulation—I need to tell you how this whole partial-density-thing is produced in the first place. The key is speed.

"Remember that the e-comets must obey a speed limit at all times. They must cross each unit of space with steadfast precision on the Admiral's clock. So, when these circulating e-comets confront more space during a part of their rotation, they circulate more slowly.

"Think of it *this* way. When a cluster of e-comets accelerates through space, it trades rotational speed for lateral speed. If taken all the way to the speed of light, then all rotation disappears, and the cluster dissolves into pure energy.

"And here's another way to think of it. Imagine you are a grandfather clock in an enchanted castle. You are magically blessed with the freedom to roam the halls, though your

© 2004 BY L.N. SMITH

freedom comes at a price. For every step you take through the castle, you divert energy from your hands to your feet. The more steps you take, the fewer minutes you record on your face.

"Contrary to what you may have heard, time does not stop at the speed of light. What does grind to a halt, however, is the churning of the e-comets in the atoms of your clocks and metabolisms.

"Well, it's almost time to restart the universe and show the creation of your planet. But before I do, there's one more detail to be addressed. This one goes out to the science-minded folks.

"You might be thinking that I've blundered—that I've missed the most critical question. Not so. I've merely saved it for last.

"How can it be that light waves in a vacuum always measure the same speed in every direction, even when the source happens to be moving? Such behavior is truly unexpected, since none of the other wave-types perform in this way. For instance, if I yell out the window of a speeding train, my sound waves will travel slower forward than backward due to the air rushing past. Shouldn't light be the same way? Shouldn't the speed of light depend on the direction of the e-comet breeze? Well it does; you just haven't realized it yet.

"The first of your scientists to look for differences in the speed of light were Albert Michelson and Edward Morley in 1887. They guessed that the earth was like a train rolling through the countryside on a calm day: the stationary ether of outer space should flow across the earth like the air across a moving train. They predicted that the speed of light would depend solely on the ether, measuring faster *with* the breeze than against it. But Michelson and Morley found no such variations in speed. It seemed that light traveled along at a constant clip in every direction, regardless of the earth's motion through space. Their oversight was a simple one. It involved the placement of their instruments.

TIME VERSUS THE CLOCK

Time is not relative: it marches steadily forward. It is *clock-speed,* alone, that is relative.

Clocks, metabolisms, and all material processes advance at a rate determined by the density of space, because the tightly bundled e-comets of matter cycle more slowly in the vicinity of other e-comets. The effect is commonly misinterpreted as a slowdown of time, when in fact it is merely the slowdown of a clock. The tiny muon particle illustrates this point perfectly.

Cosmic rays are a shower of matter from outer space that slams into the earth's upper atmosphere. These particles splinter and shred into an assortment of smaller particles as they continue their race toward the ground at nearly the speed of light. One such spin-off particle is the short-lived muon.

The muon is so unstable that it should disintegrate long before reaching the ground. But it doesn't. How does it manage to survive the trip? Scientists have reasoned that time must run slower for the high-speed muon, causing this delay in its decay. Norbert, however, suggests that the answer demands more careful wording. It is not actually time that slows for the muon; rather, it is the churning of the muon's e-comets that slows. The high-speed muon is just like the grandfather clock roaming the castle; its rotational speed has been sacrificed for lateral speed. The muon disintegrates in slow motion as it tumbles to Earth.

© 2004 BY L.N. SMITH

"If Michelson and Morley had been measuring the speed of sound from a moving train, they would have known to open the window and stick their instruments outside into the flowing air. They would have known that the air *inside* the train was stagnant and would measure the same speed for sound in every direction. But this comparison did not occur to them during their experiment with light. The two men performed their test within the 'stationary' pool of e-comets being dragged around by the earth. They failed to stick their instruments out the window and into the breeze.

"So how does one detect the flow of e-comets and prove that light's speed depends on it? Here's a test.

"From a fast-moving spacecraft flying far from planets and stars, simultaneously jettison two identical atomic clocks, sending one forward and one backward of the craft. Synchronize the clocks prior to launch, and set them to trigger a light signal back to the main craft after a specified length of time. Then, note the moment of arrival for each signal.

"If space is truly a void and has no flow, then both signals should reach the craft at exactly the same instant. But, if space *does* have a flow—an e-comet breeze—then the signals should arrive separately."

Norbert takes a long pause.

"Now, I can't deny that I'll be disappointed if there's no flow to the universe; the ethercomets of Integral Theory are so simple and so beautiful that I want them to be true. But I guess it's okay if the whole thing's a bust. What matters to me most is that my listeners gain a practical understanding of the universe.

"So please, get ready; it's time to move from design to construction. Let's pull back the pendulum and replay the big bang."

The New Look of the Atom
. . . a solar system teeming with comets   by L.N. Smith

On a television screen, a gleaming two-seated roadster revolves slowly on a turnstile. A woman, located off camera, speaks with a soothing French accent. "The E-comet Cruiser is the car of the twenty-first century. Everyone will want one. It runs forever; it never collides with another; and it has perfect cruise-control. Please join me for a test-drive."

A moment later, the video takes the perspective of a passenger looking through the windshield of an E-comet Cruiser, speeding along a two lane road. The surrounding countryside is a wasteland, barren and flat: there are no trees or houses or landmarks of any kind for as far as the eye can see. The road is empty too. The E-comet travels solo on this thin ribbon of highway that stretches to the horizon.

Eventually, strange parking lots dot the landscape, and an occasional car passes in the opposing lane. Every passing car is an E-comet Cruiser like this one, and every encounter takes place at a bend in the road, forcing a momentary downshift.

The soft voice of the guide returns. "We're entering the sprawl of development around a nearby city—the smallest city in the world, in fact. This city has a familiar name, though I'm keeping it a secret until later. Perhaps you'll guess it."

Across the forward horizon, a large expressway rises through the glare. It is a highway unlike any other I've ever seen before. The expressway runs a full story off the ground, like a long bridge, allowing each crossroad to pass freely beneath it. But that's just the first of its surprising features. This elevated freeway is also a double-decker, with six lanes on the lower level and two lanes up top. An elaborate network of entrance and exit ramps connects these levels to each other and to the ground. The structure looks like a giant serpent cast in stone, coiling its way along the plains. The most surprising feature of all, however, is that this monstrous expanse of concrete is completely void of traffic: all eight of its lanes are empty. And it seems my vehicle won't be driving on this freeway either; a barricade blocks the entrance ramp.

"The road we're cruising under is the outer beltline highway. It completely encircles the city, but it is only open to traffic when needed."

Speaking of traffic, there's more of it now on this two-lane road. And just like before, each encounter involves a curve in the pavement and a downshift of the engine.

There's a funny thing about this downshifting, though: it slows the motor but not the car. The oncoming vehicles zoom past at normal speed.

And the speedometer merely confirms this bizarre inconsistency. The needle moves across the dial, but so do the numbers. They squeeze and stretch in such a way as to make the needle always point to the same speed. Weird.

Another highway appears on the horizon.

"The road you see up ahead is the inner beltline highway. It, too, loops the city. And it marks the official city limits."

The approaching inner beltline is another completely elevated road, though it looks more normal than the last one: it has only one deck, and I can see a steady stream of cars tooling along in the near lane. Road signs point to an entrance ramp, and my E-comet Cruiser climbs onto the freeway.

As I merge with traffic, it becomes clear to me that this road is also peculiar. It has just two lanes—which seems too few for an expressway—and only one of the lanes is open at this time, meaning traffic can only flow in one direction.

My E-comet travels a short distance, to the first exit ramp, then takes the clover leaf down to another two-lane road. Traffic picks up once again, causing more turns and downshifts.

"We've entered the neighborhoods surrounding the central city. Downtown is straight ahead."

The guide had mentioned that this was the smallest city in the world; I now know what she meant. This "downtown"—rising from these "neighborhoods" of parking lots—consists of a single skyscraper.

As my E-comet closes in on the lone building, my inspection of it reveals yet another structure that is not quite what it seems. The skyscraper is not a traditional building, but rather, a parking garage. E-comet Cruisers sit along the outer railings of every level.

My approach to this singular tower is wildly indirect, looping me through the surrounding neighborhoods many times before I near the building.

Gated entrances control access to the parking deck, and every entrance is currently closed. It's easy to see why: the first and most visible level of the deck is packed with cars. There isn't even a driving lane open in the center. Cars cover every square inch of pavement.

As my E-comet gets closer to the ground-floor railing, there's something else quite astonishing. The cars inside are not actually stationary; they are moving. Every one of them is rolling slowly forward, forming a long parade of E-comets that weaves its way back and forth through the structure like a drill team in tight formation. There's no room for another

© 2004 BY L.N. SMITH

car in there. Thus, I'm surprised when the guide announces that we'll be entering the tower shortly.

As my E-comet whips alongside the railing, it suddenly turns and enters the tower through one of the gated driveways. There's an immediate and powerful downshift of the engine, though I'm surprised to see that my car is not actually slowing down. It careens into the garage of cars, but doesn't cause a collision.

The E-comets within the tower move swiftly now, matching my own car's high speed. They must have accelerated as I entered, so I check my speedometer for confirmation. Sure enough, the needle points to the same number as it always has, but now, the number is squeezed next to the zero. All of the numbers and hash marks on the dial have slid together into a slender wedge. And one look outside the parking deck reveals what has happened.

Cars blur past the tower at blinding speed, and shadow lines sweep through the structure, turning day into night and night into day in rapid succession. The slowdown I was expecting inside the tower turned out to be a speedup of everything else. I conclude that time measures more quickly outside the tower.

As my E-comet climbs higher and higher within the structure, I gain an excellent view of the city. The accelerated flow of traffic through the streets produces clear patterns, with the most dramatic ones appearing at night.

The inner beltline highway, with its single open lane, creates a constant ring of light around the city. Headlights and taillights curl from this ring at each of its many entrance and exit ramps.

Nighttime revelers filter into town from the countryside, swelling the streets with traffic. Then, all at once, an upper lane of the outer beltline highway opens to relieve the congestion, while the sole operating lane of the inner beltline closes simultaneously. Late in the night, a rush of taillights to the exit ramps causes the outer loop to close and the inner loop to reopen, as the out-of-town guests head for home.

At sunrise, a wave of commuters floods the city from the east, and traffic once again diverts from the inner to the outer expressway. The process reverses itself at five o'clock, when the commuters leave town.

On the weekends, road rallies visit the city. These packs of E-comet Cruisers arrive in large numbers, but seldom get as far as the inner beltline; they are usually turned away just before it. On rare occasions, however, a select rally gains access to the second lane of this highway for a short spin around the city before heading back into the countryside.

"The streets are quite busy, wouldn't you agree?" remarks the guide. "But this is the summer tourist season. It's not so hectic here in the wintertime, when super-cool temperatures send people packing for warmer climates. Many streets are deserted then.

"Perhaps you've noticed that the highways circling the city don't use all of their available lanes. In this city, there's rarely more than one lane open at a time. Any idea why? The answer is simple once you know the function of this parking garage. You see, this skyscraper is the traffic control tower for the city; it governs the overall flow of cars. Its limitation, however, is that it is a single tower; it can only manage a certain volume of traffic. In fact, this tower is already over-extended. It has extra floors of E-comet Cruisers to help handle the workload.

"So maybe, by now, you've guessed the name of this city. Its name is easily found in any high school chemistry book. You are touring the popular city of Hydrogen, France. It's

© 2004 BY L.N. SMITH

the smallest city in the world, as ranked by the Periodic Table of Elements. Hydrogen is a mere peanut compared to the hundred or so other atomic cities.

"Take, for example, the largest metropolis still standing: Bismuth, Germany. Its downtown has 209 such massive parking decks, and its beltline system has not two, but six looping freeways, with the outermost ring having six decks that range from two to twenty-two lanes. The city manages to keep 83 of its 182 lanes open most of the time.

"But Bismuth is not the biggest city ever built. Germany's Darmstadtium was much bigger. This mega-metropolis had 281 central towers, 7 beltlines, and 110 lanes open to traffic. But such a large city couldn't last long. Again, the problem was traffic control.

"Darmstadtium's 281 control towers were too difficult to coordinate. Each tower needed extra levels of E-comet Cruisers simply to keep things organized. The city became so bogged down in its own bureaucracy that it had to subdivide into smaller cities. The city planners called this maneuver *fission,* and watched joyfully as unneeded E-comets fled the towers and radiated out into the countryside.

"So what about tiny Hydrogen, with its overextended tower? How might this littlest city solve the problem of too many E-comet Cruisers in its downtown building? The answer is *to grow.* Hydrogen must merge with another city so that their combined towers can share the workload. The city planners call this maneuver *fusion,* and again, they enjoy the liberation of E-comets.

"So you might be wondering, what is the city of optimum size? —What city has the correct balance of towers and traffic? The answer is *the ancient Sumerian city of Iron.* Iron is the most stable city in the world, with 56 towers, 4 beltlines, and

26 lanes open to traffic. Any attempt to change this city's size through fission or fusion requires the enlistment of additional E-comets for service within the downtown skyscrapers."

Sunshine pours into my car upon reaching the top platform of the parking deck. This is my best view of the city so far.

"Take a good look down there. Hydrogen is very much like the other cities I've described to you. But there *is* something I haven't told you. Every city I've mentioned has an identical twin . . . well, *nearly* identical. These sister cities exist in England, across the channel. The difference between an English city and the others involves the flow of traffic. In England, E-comet Cruisers drive on the other side of the road. I'm sure you can see the wisdom of keeping English cities from venturing onto the mainland. If such opposing cities were ever to connect their roadways, traffic would scatter in all directions, leaving both cities abandoned and desolate.

"Please direct your attention to the horizon, where a second city is now plowing toward us from the French-Swiss border. It's a city just like this one, having one central tower surrounded by two beltline highways. It even shares the same name as ours—Hydrogen. But what we don't know about this approaching city is whether or not it is of English design. Will our two Hydrogen cities combine in an act of fusion or covalent bonding, or will they destroy each other in an act of annihilation? We'll have our answer soon enough."

Before I'm able to note the direction of the traffic circling the oncoming city, my E-comet Cruiser dips below the top deck of the tower. It descends swiftly along a steep ramp to the bottom level.

Abruptly, the earth shakes, and my cruiser spills through one of the many exits at the base of the tower. Other E-comet Cruisers depart with me.

As my E-comet speeds from the city, I note that the downtown is still standing, ruled now by a pair of towers. I note also that the inner beltline highway has not one, but two lanes open to traffic. The sign at the city limits now reads *City of Helium.*

My guide thanks me for joining her on this test-drive of the E-comet Cruiser, and we ride off into the sunset as the film credits roll.

parking towers	NUCLEUS OF AN ATOM
lanes of a beltline highway	ELECTRON ORBITAL SHELLS
a beltline lane filled with cars	ELECTRON
cars on a beltline ramp	VIRTUAL PARTICLE
nighttime revelers	HEAT
daytime commuters	LIGHT
a road rally	FREE ELECTRON
an English city	ANTIMATTER
an E-comet Cruiser	ETHERCOMET

Copyright © 2004 by L.N. Smith

All rights reserved.

ISBN-10: 0-9653316-1-x
ISBN-13: 978-0-9653316-1-6

Printed in the United States of America

This document and its author and publisher are not affiliated with or endorsed by the Walt Disney Company.

www.ingramcontent.com/pod-product-compliance
Lightning Source LLC
Chambersburg PA
CBHW032022190326
41520CB00007B/581